Ali Abakar

The 4-stroke engine: better understanding its operation

Ali Abakar

The 4-stroke engine: better understanding its operation

ScienciaScripts

Cover image: www.ingimage.com

This book is a translation from the original published under ISBN 978-620-2-27328-2.

Publisher:
Sciencia Scripts
is a trademark of
Dodo Books Indian Ocean Ltd. and OmniScriptum S.R.L publishing group

120 High Road, East Finchley, London, N2 9ED, United Kingdom
Str. Armeneasca 28/1, office 1, Chisinau MD-2012, Republic of Moldova, Europe
Printed at: see last page
ISBN: 978-620-5-78558-4

Contents

FOREWORD

This book aims to help you understand the four-stroke engine, describe and master its working principle, i.e. the four-stroke gasoline and diesel engine. This document will be of paramount importance to beginners in the field of technology. It is designed to be affordable and understandable to anyone who reads it. Also, it will be useful to students of thermal and industrial engineering. It is a document that will help the industrialist in understanding the operation of the motors around him.

This book is composed of seven main parts, the first part will be interesting because it is about the generalities of the four-stroke engines.

But the most important part of this book is part four (04) which highlights the valve functions in these revolutionary machines. It is here that we learn how the valves open and close at the right time during the different operating times of the machine.

At the end of each part, a problem whose correction can be found in the appendix is presented to you so that you can better understand the part and not be stuck. It is full of questions that you may face one day concerning the game. The wish is not to look at the correction until you have not looked for at least thirty minutes.

The book you are holding in your hands is at your disposal. You can copy it, photocopy it and remix it if necessary (modify it if there is an error) by adapting other exercises related to the parts.

In the appendix you will also find the different references of the pictures of the manual since I am not the author of these drawings (as I underlined above this manual is the fruit of a research and not of inventions) without forgetting the different corrections of the different problems and a computer program which allows to calculate in a numerical way the output of an engine and the wealth of a mixture.

My hope is that after using this manual, the operation of a four-stroke engine will be a genetic skill for you. Enjoy your reading.

Thank you

First of all, I would like to thank **Professor ALI AHMED**, Director of the University Institute of Technology of N'Gaoundéré for the framework and training he gave us during the academic year in which this study was conducted;

- **Pr Mouangue Ruben,** Head of the Department of Energy Engineering, for the supervision and the unfailing support that he offered us during this training;
- **M.EVENGA**, teacher at the IUT, for his programming course which helped me to write the program that I inserted as a complement ;
- **Mr. MAHAMAT NOUR ABAKAR,** head of the after-school and extra-curricular activities department for his moral and financial support;
- All the administrative staff of the University Institute of Technology of N'Gaoundéré ;
- All my friends and classmates

Signings

I dedicate this work to :

S My family, especially my dear mother, my dear father and my dear brothers and sisters for their moral and material support;

S To those who have never stopped encouraging me and advising me;

S the trainers of the University Institute of Technology **(I.U.T)** of the University of N'Gaoundéré

S All the people who are very dear to me from the smallest to the biggest.

PART I : GENERAL INFORMATION ON 4-STROKE ENGINES

The four (04) stroke engine was first conceived by Beau de ROCHAS in 1862, some retouching or development was done by Nikolaus Otto in 1867, pre-functioned by Gottlieb Daimler and Wilhelm Maybach in 1887, followed by the arrival of the Diesel engine in 1893. The **combustion engine** refers to any type of internal combustion engine with reciprocating or rotary pistons, diesel or spark ignition. This type of engine is generally used for the propulsion of vehicles as heavy as they are, but also in the industries as in chainsaws, generators...

Combustion engines are thermal machines emitting useful energy developed by the combustion of an oxidizer.

In these engines, as in the case of vehicles, the chemical energy from the fuel is transformed into mechanical energy, and it is this energy that makes the vehicle move. However, there is also a thermal energy, which evacuates gases that are discharged into nature through the exhaust pipes and can have an impact on the environment as these fumes contain CO_2 , NO_x and others. in particular are responsible for global warming in the world.

In order for the combustion in these engines to be sufficiently complete, the oxidizer is mixed with the air. The oxidant is admitted into the combustion chamber through the valves and its explosion depends on the type of engine. Therefore, the distinction of engines is not made by the type of fuel but by the mode of ignition, there are two types; the mode of ignition by injection (case of diesel engine) and by electric spark (case of gasoline engine).

PART II: STUDY OF THE 04-STROKE ENGINE CYCLE

1- RECIPROCATING PISTON ENGINE

There are rotary piston engines, turbine engines and reciprocating piston engines. The latter is a very common type of engine, where the air-fuel mixture is compressed, ignited and burned in a combustion chamber. The combustion process creates a high pressure on the moving part, the piston. The rotary motion of the crankshaft is obtained by transforming the translational motion of the connecting rod, piston and crankshaft. This rotary motion is then transmitted to the wheels to move the vehicle.

It is surrounded by the engine block with cylinders, cylinder head and crankcase.

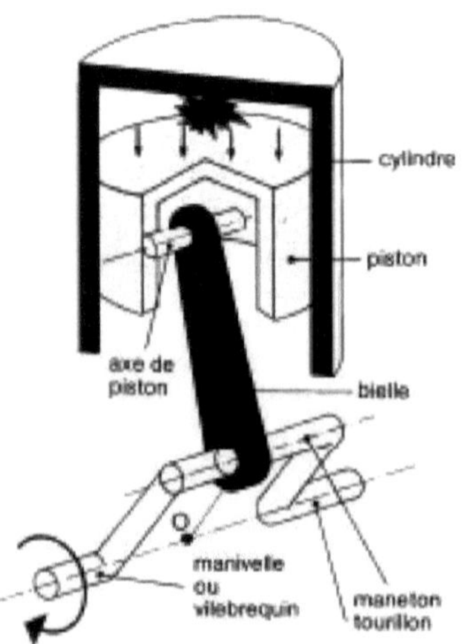

Figure 1: Transmission of the piston motion to the wheels

The 4-stroke engine: mi ,

The piston drives the swirl pin which moves the vehicle. The pistons slide in the cylinder, the cylinder head carries the valves. The opening and closing is controlled by the camshaft which is driven by the crankshaft. The turn ratio is therefore 2/1, one will make two turns while the other will only make the first turn.

2- 4-STROKE CYCLE OF RECIPROCATING PISTON ENGINES

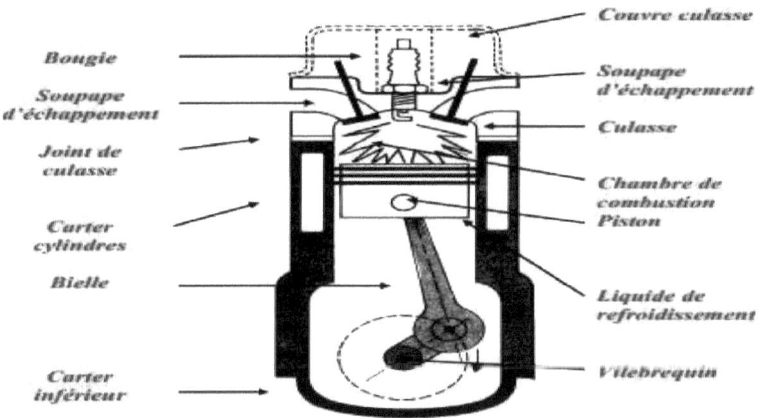

Figure 2: brief diagram of a 4-stroke engine

The diagram above shows the main elements of a 4-stroke reciprocating engine.

The main characteristics that define a single cylinder reciprocating engine are:

> **The stroke** which is the displacement of the engine between TDC and BDC (TDC: top dead center, which is the position of the piston closest to the cylinder head and BDC: bottom dead center, which is the position of the piston farthest from the cylinder head).

> **The bore** represents the internal diameter of the cylinder

> **The cylinder capacity** which is the maximum volume of the fuel mixture that can be admitted inside the cylinder and which represents the sum of the volume generated by the displacement of the piston and the volume of the combustion chamber delimited by the cylinder head above the piston at TDC.

> **The compression ratio (volumetric ratio)** which is the ratio between the displacement and the volume of the combustion chamber.

It is noted: $Rv = \frac{Va+Ve}{Ve}$

The time required to transform the calorific energy of the fuel into mechanical energy is called a **cycle.** It is important to note that each cycle corresponds to a new mixture. The engine cycle consists of two crankshaft revolutions and four piston strokes, which are the four strokes of the engine. A closed cycle corresponds to the four strokes, including the **intake of the fuel mixture, the compression of the mixture, the ignition, explosion or expansion** and **the exhaust of the burned gases.**

3- OPERATION OF THE 4-STROKE GASOLINE ENGINE

The gasoline engine in an automobile is an internal combustion engine, usually referred to as an internal combustion engine. It is widely used to move vehicles of all kinds and sizes, in the air, on water or on land. It is also used for power tools and to provide energy to fixed installations.

To do this, the four strokes of a gasoline engine are :

> **Intake:** during this time, the fuel and air mixed by the injection or carburetor are drawn into the combustion chamber by the descending piston, the valves being open.

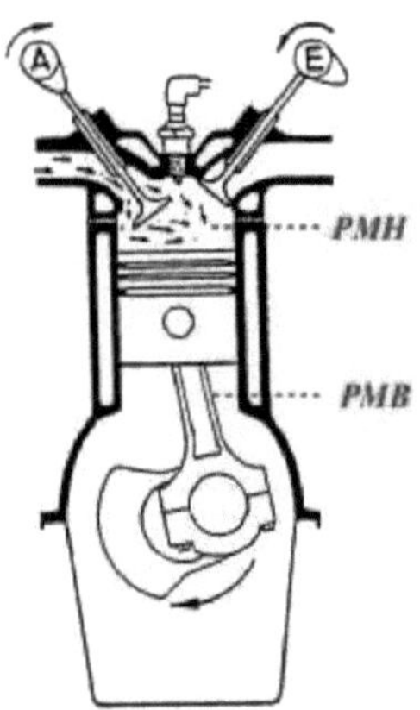

Figure 3: First stage or Admission

> **Compression:** here, the intake valves close, the piston rises and compresses the

mixture to a good pressure

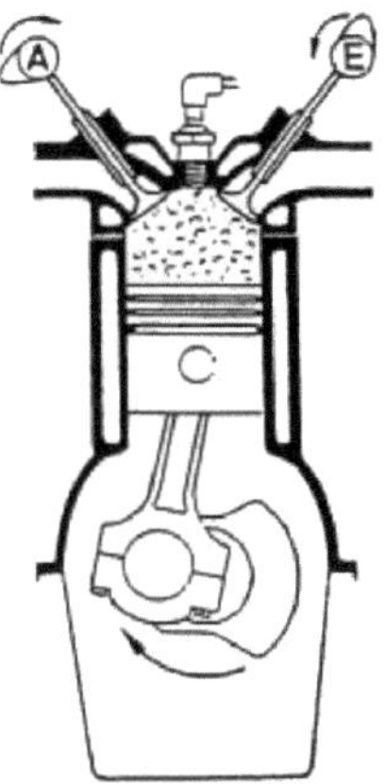

Figure 4: Second stage or Compression

> **Combustion and expansion:** the piston arriving at its top dead center will compress the gas to the maximum. The spark plug will produce a high voltage spark a few degrees before TDC is reached. This triggers a combustion of the gas which pushes the piston back.

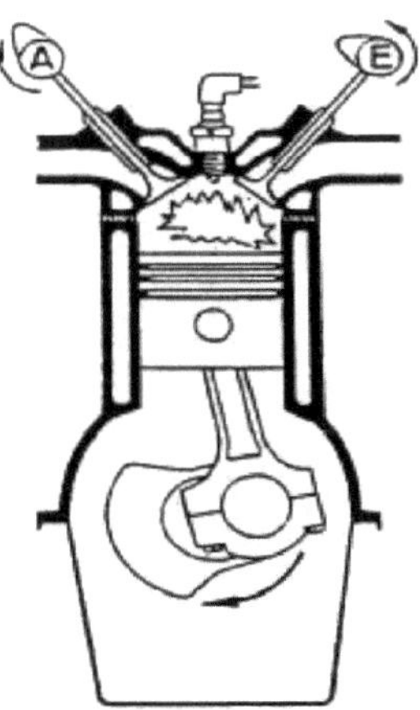

Figure 5: Third time or Explosion

> **Exhaust:** the exhaust valves will open to let out the gas which is pushed by the piston which goes up.

And so the cycle begins again.

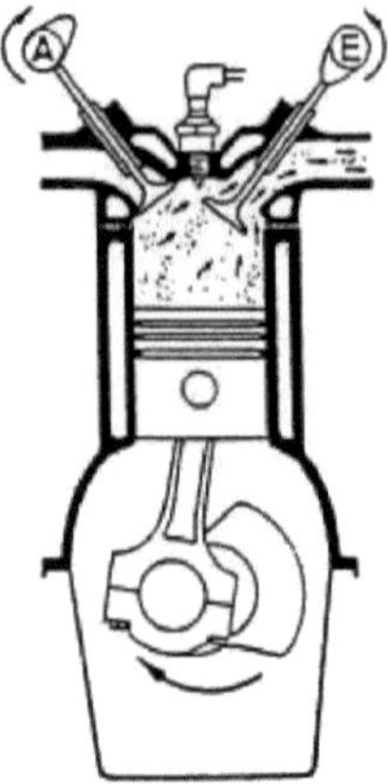

Figure 6: Fourth time or Escape

4- 4-STROKE DIESEL ENGINE OPERATION

A diesel engine is a piston engine that operates on the principle of self-ignition. The pressure and temperature in the core of such an engine reach such high levels that the fuel ignites spontaneously.

This concept differs from a gasoline engine where the ignition of the fuel mixture is triggered by the spark of the spark plugs.

For this engine we also distinguish four times

> **Inlet:** the piston moves downwards, then the inlet valve opens allowing only air to pass into the cylinder.

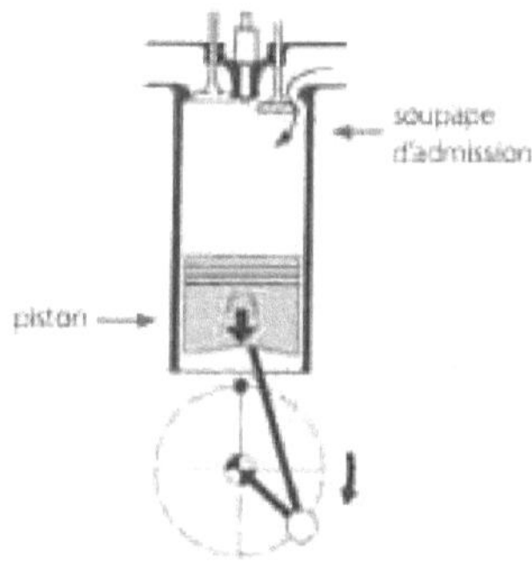

Figure 20: First stage or Admission

> **Compression:** while both valves are closed, the piston moves upwards compressing the air in the cylinder. The diesel fuel is injected into the hot air by means of an injector located in the valves. Due to the heat of the compressed air, the fuel sprayed in fine droplets ignites.

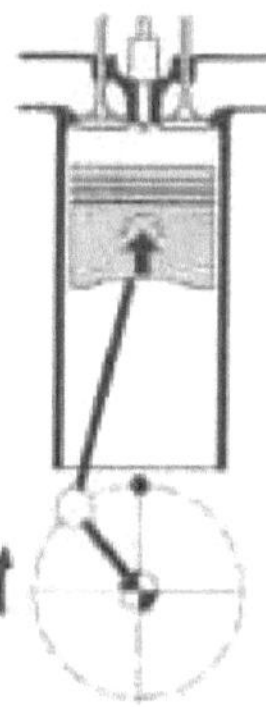

Figure 7: Second stage or compression

> **Working time:** under the effect of the pressure increase, the piston is pushed down and the energy is transmitted to the crankshaft. It is important for us to note that during this time both valves are closed.

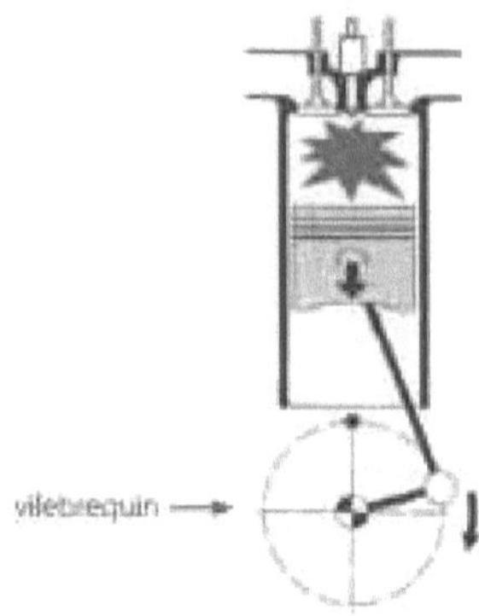

Figure 8: Third time or Explosion

> **Exhaust:** the piston moves upwards while the exhaust valve opens. During this phase, the burnt mixture is pushed outwards.

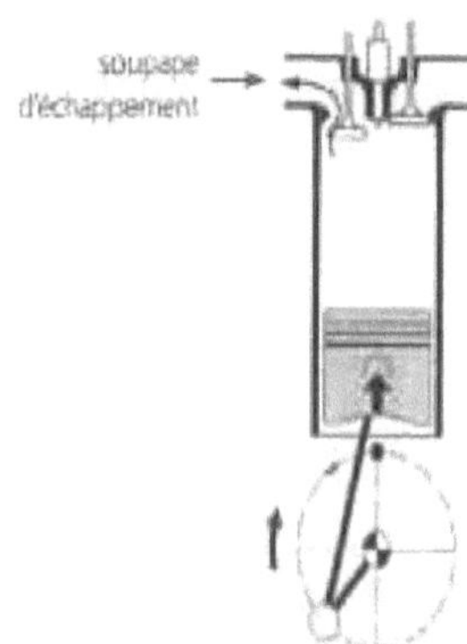

Figure 9: Fourth time or Escape

Thus the cycle begins again.

5- DIAGRAM OF THE 4-STEP CYCLE

The pressure variations in the combustion chamber are visualized by a pressure-volume (P-V) diagram as a function of the piston position.

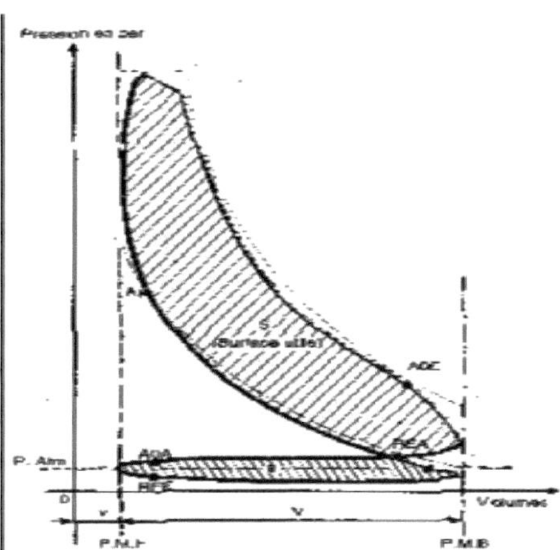

Figure 10: Diagram of the four-step cycle

- The intake between the points AOA and RFA during which there is depression. In a simpler way, the RFA point corresponds to the moment when the gas pressure is equal to the atmospheric pressure.
- The compression between the points RFA and AI during which the pressure increases due to the decrease of the cylinder volume.
- The combustion and expansion between points AI and AOE during which the pressure increases due to the increase of the gas temperature reaches its maximum some degree after TDC, then decreases due to the increase of the volume and the fall of the gas temperature.
- The exhaust between the points AOE and RFE during which the pressure continues to decrease.

On this diagram, we notice two distinct zones, **S** upper and **s** lower, corresponding respectively to the work actually done and the work absorbed by the filling and evacuation of gases.

The difference in the area of these two zones represents the available work provided by a 4-stroke cycle. On the basis of this diagram, the manufacturers determine by calculation **the average pressure indicated**. They usually also indicate the effective pressure which is slightly lower than the indicated average power because the mechanical efficiency is included.

Definition of acronyms

- **AOA:** Advance Opening Admission
- **RFA :** Return Closing Admission
- **AI :** Admission Injection
- **AOE :** Advance Opening Exhaust
- **AA:** Advance Admission
- **RFE :** Return Exhaust Closing

6- MULTI-CYLINDER ENGINE

In order to provide smooth movement of the motor vehicle, it is essential to provide a smooth torque to its output. In order to provide high power and smooth rotation, engines are often equipped with several cylinders. The four strokes of the cycle are distributed over two rotations of the crankshaft. The ignition interval is therefore on four-cylinder engines 180° on two cylinders 360° . This is achieved by an appropriate arrangement of the cylinders and crankpins of the crankshaft. The position of the cams on the camshaft in a certain sequence determines the order of function of the cylinders. The ignition system produces electric sparks in a certain order of ignition, the most common being the 1342.

Sizing in engines

Gasoline cycle

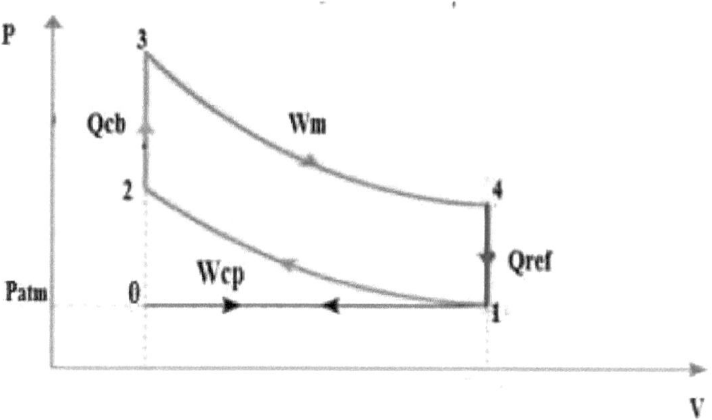

Figure *12:* Gasoline cycle

> Work indicated by the cycle: it is the work provided by the gas during a cycle

$$W_i = \int -PdV$$

To take into account the mechanical losses we introduce a mechanical efficiency η_{mec} and an effective efficiency per cycle. $W_{eff} = \eta_{mec}\, W_i$

The indicated efficiency will be $\eta_i = W_{i\ /\ Theoretical}$

> Effective yield

$$\eta_{eff} = P\ /P_{eff\ comb} \text{ with } P_{comb} = m_{comb}\ PCI$$

> Mass admitted by the cycle

$C_{ycle} = \varepsilon p\ (V - V_{12})$ with p the mass of volume of air admitted and $(V - V_{12})$ the unit displacement.

PCI (lower calorific value); it is the quantity of heat necessary so that after combustion the water contained in the fuel is found in the vapor state.

PCS (superior calorific value); it is the quantity of heat necessary so that after combustion the water contained in the fuel is in a liquid state.

> Number of cycles per second noted **x**

Let n be the number of cycles, N be the engine speed in rpm, and ω be the angular speed of the engine shaft.

X= $\omega/n'=2\pi N/60n'$ with n'=4

$P_{eff} = \omega_{eff}\, X$; $P_i = \omega_i * X$

> Average effective power is the power that would have to be applied to the piston during a cycle to obtain the same effective work. The average effective power is used to indicate the load level of the engine. For small diesel engines, the Pme is about 7 bars. For gasoline engines it varies between 8 and 12 bars.

A major interest of Pme is that it allows to compare the work done by engines of different displacements.

$Pme=W_{eff} / \gamma C$

> The effective specific conformation

$$CSE= \frac{mcomb*3600}{Peff}$$

In a similar way, we can define the performances (Pmi, Csi, Pi) which correspond to what we could measure if there were no mechanical losses due to friction and the driving of accessories.

Sme+Smf=Smi Cse.Sme=Csi.Smi

With Pmf the average friction power

> The richness, a mixture is said to be rich if there is more fuel than oxidizer

$$R = \frac{\left(\frac{qm\ combustible}{qm\ comburant}\right) reel}{\left(\frac{qm\ combustible}{qm\ comburant}\right) stoechimetrique}$$

- If R < 1 the mixture is lean

- If R = 1 the mixture is stoichiometric
- If R > 1 the mixture is rich

Stoichiometric condition

1 kg of fuel is equivalent to 15 kg of air

1 m3 of fuel is equivalent to 2.84 m3 of air.

Diesel Cycle

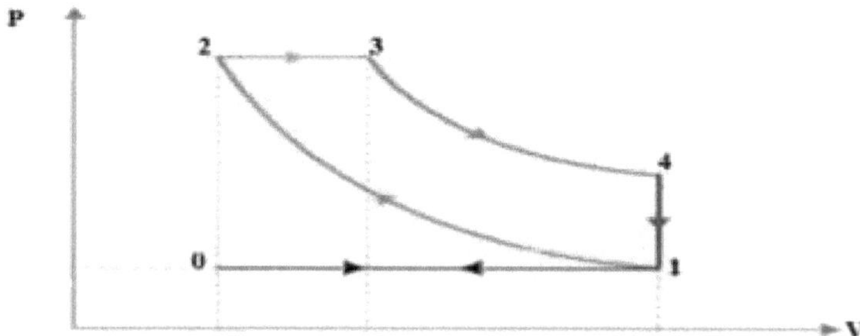

Figure 11: Diesel Cycle

$E = V / V = T_{3232}$ /Tpercentage of volumes in the combustion.

ρ volumetric pressure ratio V / V_{12}

(1-2) adiabatic reversible compression $Q_{12} = 0$; $W_c = (U_2 - U_1) = m * C_v * (T - T)_{21}$

(2-3) isobaric combustion of fuel $Q_{comb} = (h - h_{32}) = C_p (T - T);_{32}$

$W_{comb} = (U - U_{32}) - Q_{comb}$

(3-4) reversible adiabatic expansion

(5-1) air intake

(4-1-5) venting and exhaust

$\lambda = P / P_{41} = T / T_{41}$ $Q_{comb} = C\ T_{p1}^{(\gamma-1)}\ (\varepsilon-1)$

Thermal efficiency

$$\eta_{th} = \frac{W_{utile}}{Q_{comb}}$$

7- PROBLEM

A. Give the difference between a 2-stroke engine and a 4-stroke engine

B. Explain the four strokes of a gasoline engine

C. Explain the four strokes of a diesel engine

D. What do the terms AOA, RFA, AOE, RFE, AI and AA mean?

E. What is the difference between diesel and gasoline engines?

F. What is the influence of the mixture compression in the combustion chamber?

G. How is auto-ignition achieved and in what type of engine is this ignition mode found?

H. Recall the formula for the richness of an engine and the stoichiometric conditions

I. Why do we say that during the four-stroke cycle, only the third stroke is driven?

J. Interpret the P-V diagram of the following 4-stroke cycle

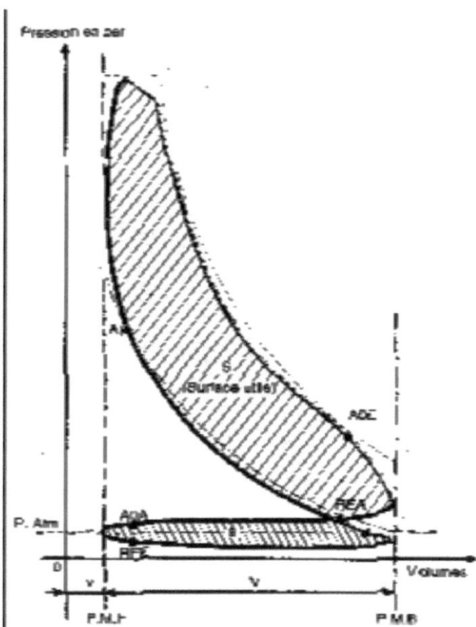
P. Atm
S
0
Volumes
v
V
P.M.B

PART III - DISTINCTION BETWEEN GASOLINE AND DIESEL ENGINES

The gasoline engine was invented by Beau de ROCHAS in 1862, this invention has undergone improvements until obtaining an engine manufactured by Rudolph DIESEL called Diesel engine in 1893. They work from different fuel including gasoline, diesel etc.. the main difference between these two engines occurs in the third time which, you know and the time of explosion.

In the gasoline engine, the third stroke or explosion is made from the small high voltage sparks that are produced by the spark plug. More explicitly, the gasoline-air mixture contained in the combustion chamber is compressed in the combustion chamber by the piston that goes up in the second stroke and the spark produced by the spark plug creates the explosion.

In the diesel engine, on the other hand, the third stroke or explosion is carried out by injecting the fuel into the combustion chamber containing the air admitted in the first stroke. The pressure generated by the rise of the piston in the second stroke compresses the air already in the combustion chamber, the injection of the fuel into the chamber is mixed with the air under high pressure and an explosion occurs.

However, there are also differences in the disadvantages and advantages of the two engines.

The advantages of gasoline engines are that they are generally more powerful in terms of horsepower, and they are also easier to adjust and tune. In addition, gasoline engines have a better price/power ratio and they burn more ecologically, and this advantage, which may be negligible, is that they do not emit enough noise.

The diesel engine has many advantages, such as low fuel consumption, low price per liter of diesel fuel, long service life, and high residual value.

Speaking of disadvantages, the major disadvantage on the gasoline engine is related to its lifespan, it is generally short, but one cannot miss the fact that the gasoline engine makes less miles per gallon than the diesel equivalents, one should also note that the residual value of the gasoline engine is low and they also have a low torque.

The diesel engine, as perfect as it seems, has also disadvantages, such as the noise it emits, the way it burns is not very ecological and also the maintenance cost is high but it is also very expensive.

PROBLEM

A-What is the difference between a gasoline engine and a diesel engine according to the mixture formation?

8- What is the difference between a diesel engine and a gasoline engine in terms of ecology?

C- Give two disadvantages of the gasoline engine

D-Give two advantages of the diesel engine

E-A fuel of Q=1200 W powers a car engine that provides a useful work of W_{utile} =800 W, calculate the thermal efficiency of the engine.

PART IV - OPERATION OF THE DISTRIBUTION

1- NEED AND STRUCTURE OF THE DISTRIBUTION

The study of the four-stroke cycle of the reciprocating piston shows that the opening and closing of the exhaust and intake valves must be done at the right time in relation to the piston position. The movement of the pistons must be rigorously synchronized with the movement of the valves. Since the piston is driven by the crankshaft, the mechanism that connects the valves to the crankshaft is called the timing mechanism. In fact, it includes :

> **The camshaft control** the first transmission that happens is rotary, it is done between the shaft and the crankshaft

> At this stage, the transmission is alternative, it is done between the cams of the camshaft and the valves themselves.

The valves are operated by the cams of the camshaft. The lift and the duration of this lift is determined by their special profile.

2- DISTRIBUTION DIAGRAM

The opening and closing of the valves is carried out taking into account the TDC and the BDC of the piston. This position is indicated by the angle of rotation of the crankshaft.

By plotting the opening and closing angles of the valves in relation to the rotation of the crankshaft and the TDC and BDC of the piston, we obtain the distribution diagram. This diagram allows to proceed to the timing of the distribution as well as to the precise control of the distribution.

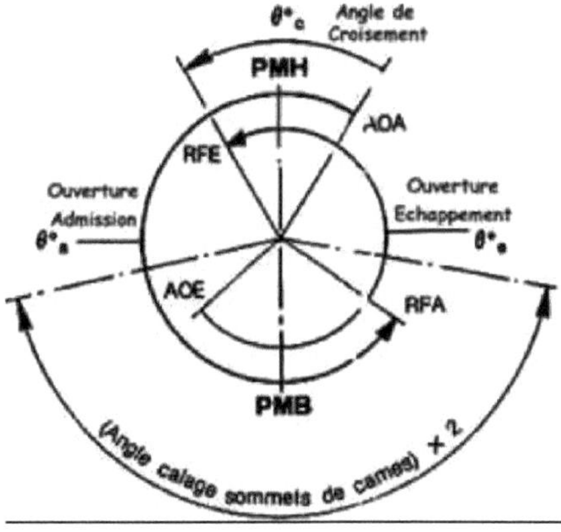

Figure 13: Distribution diagram

3- CAMSHAFT CONTROL

The camshaft is controlled with great accuracy so that the ratio of its speed to that of the crankshaft is 0.5

Every two revolutions of the crankshaft, there is a suction and a warm-up for each cycle. The camshaft makes one revolution when the crankshaft is in the second revolution. The camshaft is driven by means of helical gears, a chain or a toothed belt.

> Pinion drive; this drive is quite classical, and it takes enough space in the engine, two pinions are mounted on the ends of the camshaft and on the crankshaft, the gearing of the teeth of these pinions make the movement of the camshaft and the crankshaft uniform. An example is shown in the picture below.

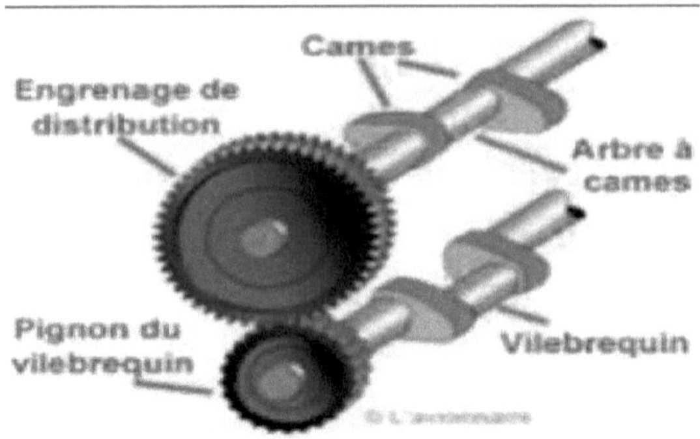

Figure 14: Pinion drive

> Chain drive; this type of drive is complex, because in spite of the sprockets mounted at the ends of the camshaft and crankshaft there is a chain connecting them. This type of control is found in engines where the crankshaft and camshaft are quite far apart and the sprockets cannot touch. This is similar to the way a bicycle works.

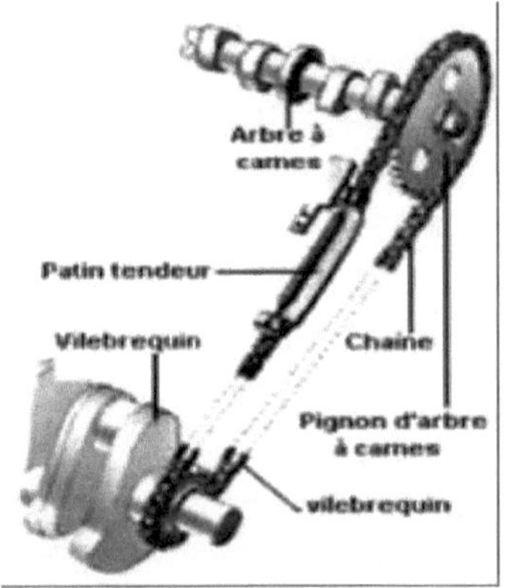

Figure 15: Chain control

> Toothed belt drive

4- VALVE CONTROL

The valves are controlled by means of mechanical elements placed differently in order to support the increase in engine speed.

> Engine with lateral camshaft and lateral valve, this first arrangement has the

advantage of its simplicity

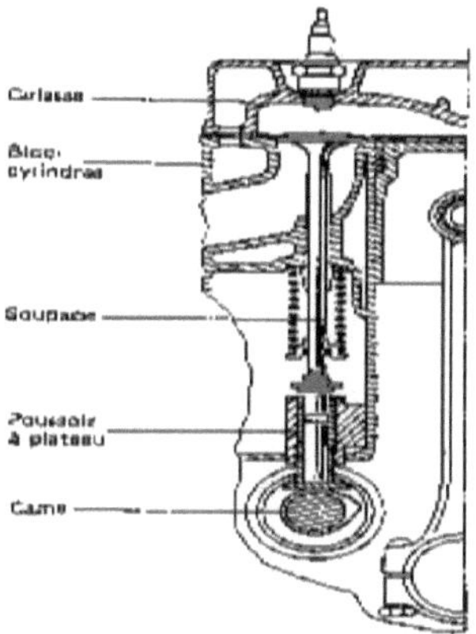

Figure 16: Motor with side camshaft and side valve

> Camshaft motor with overhead valve

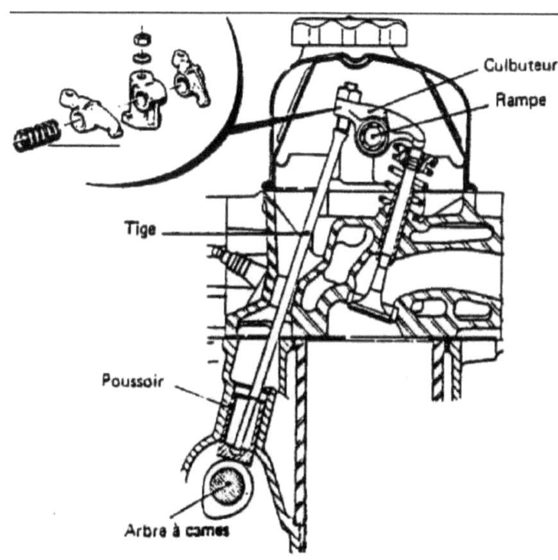

Figure 17: Motor with side camshaft and overhead valve

5- PROBLEM

A-Cite and explain two types of camshaft control.

8- What is the purpose of the distribution?

C- What are the components of distribution?

D-What is the difference between toothed belt drive and chain drive?

PART V - OPERATION OF LUBRICATION AND COOLING SYSTEMS

1- NEED AND OPERATION OF THE COOLING SYSTEM

The study of the four-stroke cycle of the reciprocating piston shows that during combustion, the parts of the engine heat up strongly, particularly the piston, the cylinder and the cylinder head. This heating has consequences such as the modification of material properties, increased risk of auto-ignition, expansion of parts and consequently a reduction of clearances, alteration of the lubricant...

In order to limit the problems related to overheating, the engines are equipped with a cooling system that conducts the heat of the engine to the atmospheric air, it is the same system applied in the chimneys, the heat is sucked by a fan and pushed back outside. This means that when you put your hand in front of a car, on the engine side, you feel like you are being burned.

The cylinders and the cylinder head are equipped with a double wall. The space between these walls is filled with a coolant. The purpose of the cooling system of the gasoline and diesel engine is to bring the engine to operating temperature as soon as possible after it is started, to keep the operating temperature of the engine constant.

The most common cooling system is the forced liquid circulation type.

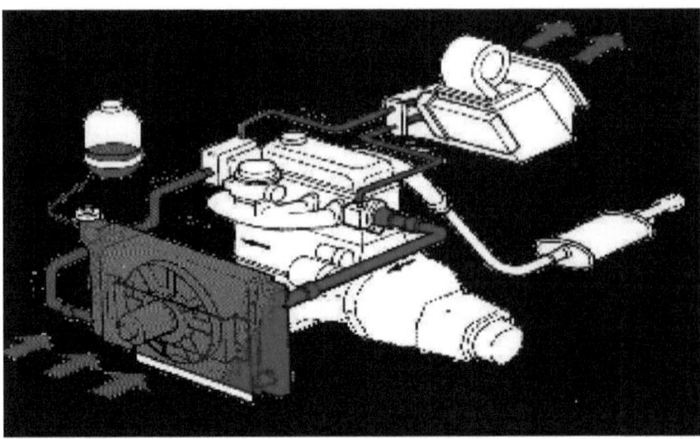

Figure 18: Engine cooling system

2- NEED AND OPERATION OF THE LUBRICATION SYSTEM

The study of the four-stroke cycle of the reciprocating piston engine shows that the different elements of the engine slide over each other, for example the piston and the cylinder.

In order to avoid dry friction, considerable heating and heavy wear of parts, the engines are equipped with a lubrication system that carries a layer of lubricant between the moving parts. The purpose of the engine lubrication system is to contribute to the thermal balance of the engine, reduce friction and avoid wear of the parts in contact, and reduce the noise emitted by the contact of the parts.

This is an example of a type of lubrication system

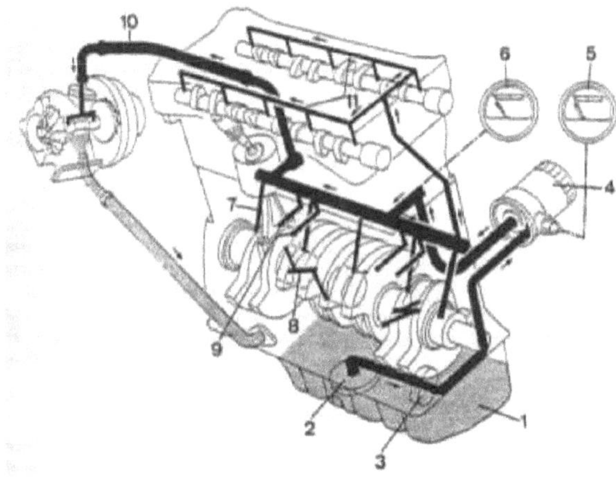

Figure 19: Motor lubrication system

1- Lower housing

2- Suction crepid

3- Oil pump

4- Oil filter

5- Pressure gauge

6- Thermometer

7- Watering cylinder

8- Crankshaft watering

9- Piston watering

3- **PROBLEM**

a- Explain the need for and operation of the cooling system

b- Explain the need for and operation of the lubrication system

PART VI- INSPECTION OF THE COMBUSTION CHAMBER SEAL

1- SEALING OF THE COMBUSTION CHAMBER

For a given engine, the force acting on the piston and subsequently the engine torque and power are a function of the pressure that reaches the gases due to combustion. The combustion itself is a function of the compression. The temperature of the fuel mixture and its homogeneity for the gasoline engine, the air temperature for the diesel engine are influenced by the compression function.

The compression pressure and the force acting on the piston depend on the absence of pressure leaks, i.e. on the tightness of the combustion chamber. Pressure leaks at the end of compression can occur in the rings between the piston and the cylinder and in the area between the valve and its seat.

The check of the pressure at the end of compression gives an overview of the tightness of the combustion chamber.

2- THE COMPRESSIOMETER

They are generally equipped with a recorder, it is a device that measures the pressure in the combustion chamber at the end of compression, it is measured in bar. The recorder compressiometer is available in two versions, the gasoline engine version and the diesel engine version.

3- PRESSURE CONTROL

It consists of making comparative measurements between the different cylinders of the engines in order to establish a diagnosis. The tightness of the combustion chambers is normal if the pressure values are above an admitted minimum and the difference between

Each cylinder does not exceed 1.5 bar for a gasoline engine and 3 bar for a diesel engine.

4- PROBLEM

The effective power of volume V=2l and at N=4200 rpm and 52 kW for a four stroke engine

a- Determine the average effective power

b- Explain the operation of the pressure control

c- Explain how the combustion chamber is sealed

d- What is a compressiometer?

Reference and Bibliography

✓ Figures 1 and 2 were taken from the book BPT_sciences- Industrielles-B_2008_TSI

✓ Figures 3, 4, 5 and 6 were taken from the book BPT_sciences-Industrielles-B_2008_TSI

✓ Figures 7, 8, 9, 20 have been taken from the book MECA-moteursdiesel_for_web

✓ Figure 10 was taken from Google at the link https://www.google.cm/search?q=diagram+of+the+cycle+at+temps+pv&client=ms-android-orange-cm-revc&prmd=ivn&source=inms&tbm=isch&sa=X&ved=0ahUKEwiqrLjXy6LYAhXSLIKHaW1DusQ_AUICSgB&biw=320&bih=454%imgrc=B8MJjTh8xMbz9m

✓ Figure 11 was taken from https://www.bing.com/images/search?view=detailV2&ccid=kia9CaXJ&id=E5114E6E5F9E6CB1CE075B5B6758A9E7FE6DD9E7&thid=OIP.kia9CaXJ0-MeRD7u-aprDQHaHa&q=cycle+de+Diesel&simid=608004690640699924&selectedIndex=4&ajaxhist=0

✓ Figure 12 was taken from https://www.bing.com/images/search?q=cycle%20%C3%A0%20essence&qs=n&form=QBIDMH&sp=-1&pq=cycle%20%C3%A0%20essence&sc=8-15&sk=&cvid=CFA253C0D46642CF9E10E400A0EA8DAA

✓ Figure 13 was taken from https://www.bing.com/images/search?view=detailV2&ccid=0flgbj8k&id=1262C5FC8A892D005DEE976A7CC02C4F22FACF0

4&thid=OIP.0flgbj8kUjs7-48etQCHGwHaFj&q=diagramme+de+distribution+dans+un + moteur&simid=608029588517291593&selectedIndex=1&ajaxhist =0

✓ Figure 14 was taken from https://www.bing.com/images/search?view=detailV2&ccid=xs8Ifi7T&id=213D9803D43C1D24FEA2F7EC4A421E2B52ED08E1&thid=OIP.xs8Ifi7TRgow-QtmkEAQLAEsCs&q=commande+par+pignon+du+vilebrequin+et+de+l%27arbre+%c3%a0+came&simid=608019881871934809&selectedIndex=4&ajaxhist=0

✓ Figure 15 was taken from https://www.bing.com/images/search?q=commande%20par%20chaine%20du%20vilebrequin%20et%20de%20l'arbre%20%C3%A0%20came&qs=n&form=QBIDMH&sp=-1&pq=commande%20par%20chame%20du%20vilebrequm%20et%20de%20l'arbre%20%C3%A0%20came&sc=0-55&sk=&cvid=A8EFBCE3E5D541F9BA3F6C54D5A65684

✓ Figure 16 was taken from https://www.bing.com/images/search?q=Moteur%20%C3%A0%20arbre%20%C3%A0%20came%20lat%C3%A9ral%20et%20soupape%20lat%C3%A9rale&qs=n&form=QBIRMH&sp=-1&pq=moteur%20%C3%A0%20arbre%20%C3%A0%20came%20lat%C3%A9ral%20et%20soupape%20lat%C3%A9rale&sc =0-51&sk=&cvid=FE7AF05CC5934D4292FFC007CECA82DE

✓ Figure 17 was taken from https://www.bing.com/images/search?view=detailV2&ccid=4d3UQbkn&id=80806CBD326625B8EE3D2498553E3DB5B41B8412&thid=OIP.4d3UQbkniv1-Y7KMPRIFYgHaFj&q=Moteur+%c3%a0+arbre+%c3%a0+cam

e+lat%c3%a9rale+et+soupape+%c3%a0+t%c3%aate&simid=607997483676270744&selectedIndex=17&ajaxhist=0

✓ Figure 18 was taken from https://www.bing.com/images/search?view=detailV2&ccid=gvN2O8pG&id=68076DAE9FC15D82FD4EC7003DAF5E69B9A20349&thid=OIP.gvN2O8pGKP6YU-4RF19f2wHaE5&q=%3a+Syst%c3%a8me+de+refroidissement+des+moteurs&simid=608016145299214194&selectedIndex=14 &ajaxhist=0

✓ Figure 19 was taken fromhttps://www.bing.com/images/search?view=detailV2&ccid=TXOx8qmP&id=3FC7A3E78C8FF9A941A0A7A8E30A93B80D92CE30&thid=OIP.TXOx8qmPMf8ERx1RuFG8jAHaFU&q=Syst%c3%a8me+de+graissage+des+moteurs&simid=608035077522850588&mode=overlay&first=1

Correction of the exercises

CORRECTION PROBLEM PART II

A-The difference between a two-stroke engine and a four-stroke engine is in the valves, that is to say that on a two-stroke engine we don't have valves and while we have them on four-stroke engines.

B- **Intake:** during this time, the fuel and the air mixed by the injection or the carburetor are sucked into the combustion chamber by the descending piston, the valves being open.

Compression: here, the intake valves close, the piston rises and compresses the mixture to a good pressure. **Combustion and expansion:** the piston reaching its top dead center will compress the gas to the maximum. The spark plug will produce a high voltage spark a few degrees before TDC is reached. This triggers a combustion of the gas which pushes the piston back.

Exhaust: the exhaust valves will open to let out the gas which is pushed by the piston which goes up.

And so the cycle begins again.

C- **Intake:** the piston moves downwards, then the intake valve opens letting only air into the cylinder.

Compression: while both valves are closed, the piston moves upwards compressing the air in the cylinder. The diesel fuel is injected into the hot air through an injector located in the valves. Due to the heat of the compressed air, the fuel sprayed in fine droplets ignites.

Working time: under the effect of the pressure increase, the piston is pushed down and the energy is transmitted to the crankshaft. It is important for us to note that during this time both valves are closed.

Exhaust: The piston moves upwards while the exhaust valve opens. During this phase, the burnt mixture is pushed outwards.

D

J **AOA:** Advance Opening Admission

J **RFA :** Return Closing Admission

J **AI :** Admission Injection

J **AOE :** Advance Exhaust Opening

J ***AA*:** Advance Admission

J **RFE :** Return Exhaust Closing

E- The difference between the gasoline engine and the diesel engine lies in the ignition mode of these engines; in the gasoline engine it is small electric sparks produced by the spark plug that trigger the explosion, while in the diesel engine the explosion is triggered by injecting the fuel into the combustion chamber, and also at the time of intake; In the gasoline engine, the air-fuel mixture from the carburetor is admitted into the combustion chamber, whereas in the diesel engine, only air is admitted at the moment of admission, the fuel will be injected later.

F- The influence of the compression in the combustion chamber is that it is thanks to this compression that there is an explosion giving rise to the third stroke.

G-The auto-ignition is carried out in the combustion chamber of a diesel engine, it is the result of the injection of the fuel in the combustion chamber containing the compressed air.

H-The wealth formula is

$$R = \frac{\left(\frac{qm\ combustible}{qm\ comburant}\right) reel}{\left(\frac{qm\ combustible}{qm\ comburant}\right) stoechimetrique}$$

The stoichiometric conditions are

1 kg of fuel is equivalent to 15 kg of air

1 m3 of fuel is equivalent to 2.84 m3 of air.

I- The third beat is said to be the engine beat because it is at this beat that the engine is properly started due to combustion.

J- The intake between the points AOA and RFA during which there is depression. In a simpler way, the RFA point corresponds to the instant when the gas pressure is equal to the atmospheric pressure.

The compression between the points RFA and AI during which the pressure increases due to the decrease of the cylinder volume.

The combustion and expansion between points AI and AOE during which the pressure increases due to the increase of the gas temperature reaches its maximum some degree after TDC, then decreases due to the increase of the volume and the fall of the gas temperature.

The exhaust between the points AOE and RFE during which the pressure continues to decrease.

CORRECTION PROBLEM PART III

A-The difference between a gasoline engine and a diesel engine according to the formation of the mixture is that in a gasoline engine the mixture is carried out in the carburetor while for the diesel engine the mixture is carried out in the combustion chamber by injection of the fuel.

B- Concerning ecology, the combustion in the gasoline engine is very ecological which is not the case in a diesel engine

C- The gasoline engine has several disadvantages but we will retain its very short life and its residual value very

low.

D- The diesel engine has many advantages, but two of them are its long life and the price per liter of its fuel.

E- Let's calculate the thermal efficiency of the engine

$$\eta_{th} = \frac{W_{utile}}{Q_{comb}} = \frac{\mathbf{800}}{\mathbf{1200}} = \mathbf{0,66}$$

$$\eta_{th} = \mathbf{66\%}$$

CORRECTION PROBLEM PART IV

A- **Chain drive**; this type of drive is complex, because in addition to the sprockets mounted at the ends of the camshaft and crankshaft there is a chain that connects them.

- **Pinion drive**; this drive is quite classical, and it takes enough space in the engine, two pinions are mounted on the ends of the camshaft and on the crankshaft, the gearing of the teeth of these pinions uniform the movement of the camshaft and crankshaft

B- The purpose of valve timing is mainly to synchronize the opening and closing of the valves according to the rise and fall of the piston.

C- The main components of an engine's timing system are the crankshaft, camshaft, piston and valves

D-The difference between chain drive and toothed belt drive is that for chain drive, at the ends of the camshaft and crankshaft are fixed sprockets and are connected by a chain while for toothed belt drive, are fixed at the ends of the belt wheels and are connected by a toothed belt. The transmission principle remains the same.

CORRECTION PROBLEM PART V

A-The cooling of an engine is necessary because the study of the four-stroke cycle has shown that during combustion the parts of the engine such as the piston, the cylinder head, the cylinder heat up strongly.

To overcome this problem, engine designers have equipped their equipment with a fairly perfect cooling system. They sometimes use both a fan that sucks heat from the engine parts and pushes it out and a liquid filled in the space between the cylinder head and the cylinder.

B- The lubrication of the engine is necessary because during the movement of the parts in the engine like the piston there is friction and this friction can cause a wear of the parts and often an increase of the play between the two parts, and if there is an increase of the play between these two parts then there is no more sealing of the combustion chamber.

The lubrication of the engine is ensured by a rather simple principle. The lubricant is contained in the lower crankcase, which is conveyed to the elements to be lubricated by pipes. A suction valve sucks the oil from the lower housing into the pipe and a pump introduces the lubricant into an oil filter, from which it is drawn cleanly and sprayed onto the parts to be lubricated.

CORRECTION PROBLEM PART VI

a- Let's calculate the effective power

With Weff=52 kW, C=2l, 4200

Pme=5200/4200*2=6,19 W

Pme=6,19 W

b- The pressure control consists in carrying out comparative measurements between the different cylinders of the engines in order to establish a diagnosis. The tightness of the combustion chambers is normal if the pressure values are above an admitted minimum and the difference between each cylinder does not exceed 1.5 bars for a gasoline engine and 3 bars for a diesel engine.

c- For a given engine, the force acting on the piston and subsequently the engine torque and power are a function of the pressure that reaches the gases due to combustion. The combustion itself is a function of the compression. The temperature of the fuel mixture and its homogeneity for the gasoline engine, the air temperature for the diesel engine are influenced by the compression function.

The compression pressure and the force acting on the piston depend on the absence of pressure leaks, i.e. on the tightness of the combustion chamber. Pressure leaks at the end of compression can occur in the rings between the piston and the cylinder and in the area between the valve and its seat.

The check of the pressure at the end of compression gives an overview of the tightness of the combustion chamber.

d- it is a device that measures the pressure in the combustion chamber at the end of compression, it is measured in bar. The recording compressiometer is available in two versions, the gasoline engine version and the diesel engine version.

Complement to the book

This part is just a complement of the book, it is a program that I wrote, it will just allow to calculate numerically the thermal efficiency of an engine knowing that you already have the useful work and the amount of heat of the fuel, also the richness of a mixture knowing that you have all the data, that is to say the mass flow rates of the real and stoichiometric fuel and the mass flow rates of the real and stoichiometric oxidizer, it will also tell you for this case if the mixture is rich or lean.

You will only have to copy and paste it into a programming program, you must open the program under console application and the language is c.

The program

```
double choice, N, Wutile, Qcomb, Qmcr, Qmcmbr, Qmcs, Qmcmbs, R; printf("WELCOME TO MY APPLICATION");

system("pause");

printf("This application will allow you to calculate the thermal efficiency of an engine for which you know the useful energy and the heat of combustioni");

system("pause");

printf("It will also allow you to calculate the richness of a mixture and it will tell you at the same time if the mixture is rich or not, knowing of course all the possible coordinates");

printf("please choose the operation to perform");

printf("type 1 to calculate the yield");

printf("type 2 to calculate wealth"); scanf("%d",&choice);

if (choice==1) { printf("you have chosen to calculate the
```

```
yield"); printf("please follow the instructions");
system("pause");
printf("please enter the value of useful energy"); scanf("%d",&Wutile);
printf("please enter the value of combustion heati"); scanf("%d",&Qcomb);
system("pause");
N=((Wutile)/(Qcomb));
printf("the output of the motor is therefore:\n");
printf("\t\t N=%d", N);
} else if (choice==2) {
printf("you have chosen to calculate wealth"); printf("please follow the instructions");
system("pause");
printf("please enter the value of the fuel mass flow rate"); scanf("%d",&Qmcr);
printf("please enter the value of the mass flow rate of the combustor reel"); scanf("%d" ,&Qmcmbr) ;
printf("please enter the value of the fuel mass flow rate stoichiometric");
scanf("%d",&Qmcs);
printf("please enter the value of the mass flow rate of the stoichiometric oxidizer");
scanf("%d",&Qmcmbs);
R=(((Qmcr)/(Qmcmbr))/((Qmcs)/(Qmcmbs))); printf("the wealth is therefore:n");
```

```
printf("R=%d", R);
if (R==1)
{ printf("And your mixture is stoichiometric");
} else if (R<1) {
printf("And your mixture is poor");
} else if (R>1) {
printf("And your mixture is rich");
}
}
system("pause");
printf("           **\n");
printf("* THANK YOU FOR YOUR VISIT        ");
printf("*
```

Printed by Books on Demand GmbH, Norderstedt / Germany